Turns Out you Can Grow Money
The Basics of Value-added Agriculture

By
Darla Noble

Mendon Cottage Books

JD-Biz Publishing

All Rights Reserved.
No part of this publication may be reproduced in any form or by any means, including scanning, photocopying, or otherwise without prior written permission from JD-Biz Corp

Copyright © 2014. All Images Licensed by Fotolia and 123RF.

Table of Contents

Introduction	4
Chapter 1: You can Have your Hobby and Make Money, Too	5
Chapter 2: The Rules of the Game	9
Chapter 3: How to Get Started	14
Chapter 4: Let's Brainstorm	18
Chapter 5: Let's Talk Business	30
Chapter 6: Helpful Resources	36
Conclusion	37
Author Bio	38

Introduction

Nearly fifteen years ago, co-author, Darla Noble, had some free time at an agricultural/farming expo she was participating in. As she meandered through the other exhibits, she happened onto a short seminar that was just beginning. The name of the seminar, *Making Value-added Agriculture Work for You,* intrigued her because she wasn't quite sure what it was.

Within minutes of taking her seat, however, Darla knew she wanted to hear everything Joan Benjamin had to say. The concept of value-added agriculture (VAA) is quite simple, really. It means using what you grow for more than one purpose; increasing its value and *your* earning potential.

Darla knew almost immediately that value-added agriculture was something she could do to make her family's farm even more profitable. So that's exactly what she did. And guess what…so can you!

Chapter 1: You can Have your Hobby and Make Money, Too

Growing plants/flowers, fruits, veggies or even livestock for a hobby is enjoyable and rewarding. But have you ever thought about turning this hobby into extra income? Have you ever thought of turning your hobby-farming venture into a small business?

If not, that needs to change starting right now. It's time you consider taking your hobby to the next level.

Extra income from you gardening hobby or small farm

Now before you start rattling off excuses like…

"I can't."
"I'm not a farmer, so value-added agriculture isn't an option for me."
"I don't know anything about running a business."
"I don't even know what value-added agriculture is, so how am I supposed to do it?"

Humor me. Give me a few minutes to 'argue' against your protests before you make a decision. What do you have to lose? Absolutely nothing but a few minutes of your time, that's what.

"I can't." How do you know you can't? Have you tried? Saying you can't do something before you've even tried, or at least considered it, doesn't make much sense.

"I'm not a farmer, so value-added agriculture isn't an option for me." Farming isn't limited to those who plow hundreds of acres, care for large numbers of animals or grow acres and acres of produce. If you grow something people can eat or use, you have the ability and potential to be considered an agricultural producer. I know several people who make a decent income in the 'world' of value-added agriculture who live on less than an acre of land.

You can grow your income by growing for added value

"I don't know anything about running a business." This reason is somewhat valid, but only until you read the rest of the book. Chapter 5 covers the business aspects of value-added agriculture—everything from permits to taxes and money management to marketing.

"I don't even know what value-added agriculture is, so how am I supposed to do it?" Value-added agriculture is just a fancy way of saying you stretch a products potential by enhancing its saleability. For example: someone who grows strawberries could sell a portion of their crop as fresh berries, but use the other portion of the crop to make strawberry jelly or syrup/topping to be sold post-harvest season. This way the berry grower extends their income potential beyond a few weeks of the year and increases their marketing potential (customer base) tremendously. See? That's not so hard, is it? No, it's not. It all comes down to being focused and letting yourself think outside the box.

Strawberry syrup is a product of value-added agriculture

Are you beginning to have a better understanding of what value-added agriculture is? Are you beginning to believe that you can turn your hobby and passion for growing things into a hobby that pays? Then let's get started.

Chapter 2: The Rules of the Game

Adapting your gardening plan to a value-added plan isn't all that difficult once you decide what you are going to grow and what you are growing it for. But when you grow for profit, there are a few rules of the game you need to adhere to.

From a business perspective, the rules you'll be dealing with pertain to permits, taxes and marketing; things we will discuss in greater detail in chapter five. But there are other 'rules' that are less technical and more practical in nature. Let's take a look at some of these and how you can avoid problems by playing by the rules of VAA.

One: Do what you love so you will love what you do. If you enjoy growing flowers but switch to vegetables because you think there's more money to be made by growing and adding value to vegetables (even though you don't enjoy harvesting them), you aren't going to be happy.

And chances are if you aren't happy with what you are growing it will show in more ways than one—leaving you disappointed and floundering. Turning your hobby into a value-added agriculture project or venture does make it somewhat of a job, but that doesn't mean it should be drudgery.

Growing flowers can be profitable

Two: Have a plan. It's not enough to be an avid gardener who decided to grow a lot more of something. It's not even enough to grow that something better than anyone around. You have to have a plan for adding the value to what you grow. Otherwise all you are going to have is way more than you need of….
In order to add value to what you grow for the purpose of generating income you will need to:

Decide how you are going to add value to what you grow. The possibilities are endless, so we won't take the time to get into that now. We'll save our brainstorming session for chapter four.

Flowers or…

Vegetables or…

The possibilities are nearly endless.

Research your idea(s). Are they viable; meaning, are these ideas something people will actually want to purchase?

Hone your skills in adding value to what you grow (if you haven't already) in order to make your value-added product(s) truly marketable. Just because you make jelly or beeswax ornaments doesn't mean you make jelly or beeswax ornaments people will buy.

Be able to add value in a cost-effective manner have sources for generating that income. In other words, you need to add value, market and sell without going in the red for it to be a true value-added venture.

Know where, how and to who you will sell to. It's best to start small and go from there rather taking on more than you can handle and end up feeling overwhelmed.

On-site greenhouse sales

Starting small may mean selling at a local farmer's market and craft show or two. These are great ways to build a strong local following and reputation. Selling online (more on that later) is also a great way to start small.

Once you are comfortable with this you can grow from there. And as long as you are having fun and making money, you can grow as much as you are willing to grow.

Determine how much of yourself you are willing to give to this project. If gardening is involved are you ready, willing and able to spend more time weeding, planting and harvesting?

Have a plan—an actual written-out business plan, so to speak. You may never need it in order to obtain financing for your project, but it will serve to keep you focused and help you evaluate your progress.

Taking your growing/producing hobby to the level of being a bit more than just a hobby isn't something you can do on a whim if you have any expectations of making it a positive and enjoyable venture. Like anything you do, if it's worth doing, it's worth doing right.

Chapter 3: How to Get Started

"The day I wandered into Joan's seminar, I was looking for a way to kill some time. I wasn't looking for anything else to do," Darla says, "but something just clicked. My head was instantly filled with ideas of what I could do, how I could do it and how to make it a success."

It all starts with an idea or two…or three.

Value-added products using old family recipes

For the sake of 'argument' we are going to assume you have decided to go the value-added route. Now it's highly unlikely that you did so without having at least a couple of ideas in mind on how you plan to go about adding value to what you are already raising/producing, so let's take that first important step in getting started; the step called deciding if my idea is viable.
To help you decide we're going to go back to the example I gave in chapter one on growing strawberries the value-added way.

Let's say you decide to hold back part of your crop in order to make strawberry jelly. Before you start cooking, though, you need to make sure there's a market for the jelly.

So...remembering what I said about starting small, you head out to some of the local and regional farmer's markets to scout out the competition. In doing so you discover there are already eight people in a two county area offering strawberry jelly at these farmer's markets. Now I know eight doesn't sound like a lot, but it is when you are talking about selling to people strolling up and down deciding what to buy. To them, strawberry jelly is strawberry jelly and they are going to buy whichever vendor is cheapest and most convenient to buy from.

What should you do? You have two choices in this situation: 1) do something else besides jelly; something unique and different from other vendors or 2) you make the jelly, but make sure it stands head and shoulders above the rest for a *specific* reason. Oh, and make sure that reason is the focus of your marketing.

For example...

Your strawberry jelly has to different...special...unique. Now before you give yourself the nod of approval because you are *sure* your jelly is special and unique because it tastes better than the others. Everyone thinks theirs is the best or else they wouldn't be selling it. You have to make yours the best by making it different. Instead of strawberry jelly, make strawberry-basil jelly, strawberry-rosemary jelly or strawberry jam with thinly sliced strawberries in it. In making it unique, you have something to focus on. Instead of selling strawberry jelly, you are selling strawberry-*basil* (or whatever) jelly; the jelly that is one-of-a-kind.

Your other choice is to forget about making jelly and making something else instead. You could make strawberry tarts. You could dry the berries, mix with loose black tea and sell strawberry tea (loose) by the oz. or by making tea bags. You could puree the berries and make fruit leather. Or you could even mix crushed berries with sugar for an exfoliating scrub.

Any of these ideas is a possibility. You would just need to decide with one or ones will work for you; which one or ones you could do

WELL, which one or ones would be cost-effective and which one or ones have the greatest market potential.

Once you have your have decided on the best way to add value to your product the next thing you need to do is take the necessary steps to make sure you have the capability to grow/produce what you need.

?

Do you have the option of doing a little less mowing for a lot more growing? Can you convert some of your yard into another or larger garden area? If so, then go to it! If it's more garden space you need but don't have the option of expanding 'out' there are plenty of inexpensive options for vertical growing or growing in containers. You'll find some great tutorials on how to garden in small spaced on Pintrest or websites devoted to small-space gardening.
Or…depending on what you are growing and where you live, you might also think about investing in a small greenhouse. NOTE: By small, I mean less than $500.

Value-added grass: a profitable sheep farm on as little as three acres

Maybe you don't need any more space than what you've already got. Maybe you just need to add another hive or two of bees or a lot of elbow grease turn that old barn into a wedding chapel. Whatever it is you need to get going, make sure it's within reach WITHOUT putting any financial burdens on you.

I'm not saying you won't make any initial investments. To do so would be unrealistic. I am, however, telling you to keep them to a minimum. Remember…start small, grow slow and never get to the point when it's not fun anymore.

Okay, so now you have a viable idea or two, you know how and where you are going to market your value-added products and you have what you need to make things happen. What's left?

What's left is to enjoy what you do, make some money doing it and becoming part of the hardest-working, genuine community of people there is…the agricultural community.

Chapter 4: Let's Brainstorm

This is, to me, the best or most fun part of the entire book. This is where I get to brainstorm with you; filling your head with all sorts of ideas on how you can be a value-added agriculturist.

The first thing I'm going to tell you is what not to grow: tomatoes, cucumbers, zucchini, green beans…you know--the standard garden produce.

Garden-fresh tomatoes

These things are a dime a dozen at most farmer's markets so unless you have something *really unusual* that will add value to them, let's think a bit more outside the box.
That being said, there are things you can make from standard garden produce that would be considered 'really unusual' and value-added worthy. These include:

Teaching others to can using a pressure cooker.

Pressure cooker

A secret family recipe for tomato-based relishes or sauces such as a chow-chow or relish made from everything in the garden that uses a unique blend of spices you won't find anywhere else.

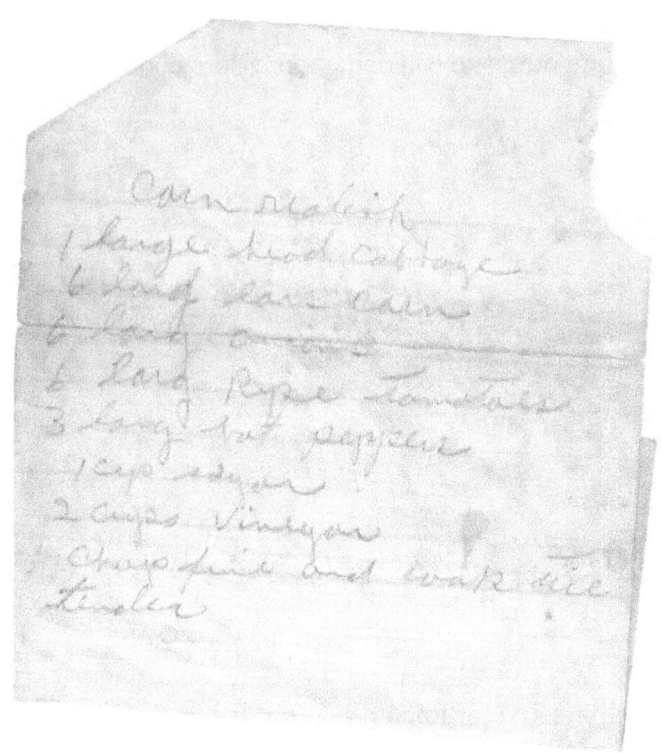

Old family recipe for garden relish

Now let's move on to ideas that will hopefully inspire you and ignite your entrepreneurial spirit. All of the following will give you both the expected or ordinary product along with some value-added options to consider.

Zinnias, Asters, Straw flowers, Roses, Daisies and other wildflowers: Sell fresh-cut flowers and seeds AND dried flower arrangements, pressed flowers for wall hangings, bookmarks, greeting cards and other home décor.

Lavender, Mint, Lemon or Cinnamon basil, Sage, Anise and other herbs: Sell plants and herbal container gardens AND dried herbs for cooking and crafts, sell to soap makers, add to tea to sell as loose tea or in tea bags you make, or make your own lotions or home spa treatments or even herbal jellies.

Mint for teas, lotions and cooking

Your small farm: Sell eggs, home-made butter and livestock for meat (delivered to the processer by you) and flock-building purposes AND give farm tours so others can see how the animals live/grow, take family pictures or children's pictures for spring or fall with a baby chick or lamb, create a child's birthday party experience basing the games on farm chores or sell manure for fertilizer.

Farm tours are a great way to add value to your farm

Succulents and House plants: Sell individually AND sell in container arrangements, teach classes on plant care, create fairy gardens, grow/sell plants that clean the air and help allergies, provide plants to be sold in gift shops (including those in hospitals) or sell/rent plants to offices.

Bees: Sell raw honey and beeswax AND flavored honeys, beeswax ornaments, beeswax lip balm, the bees themselves and offer bee-keeping classes.

Honey bees

Garden space: If you have the space but not the time nor energy to enlarge your garden, add value to your land by renting out space for others to grow their own food. Prepare the garden plot mark it off into smaller plots and rent to those who couldn't otherwise grow their own produce.

Small garden plots for rent

Empty buildings or a picturesque setting: Do you have a solid old barn or other sort of outbuilding? Is your landscaping the envy of everyone? Are your flowers beds pristine, colorful and park-like? Add value to your resources by renting them out to photographers for family, senior and engagement pictures, for wedding ceremonies and other celebrations.

Fruit trees and Berries: Sell the fruit you-pick fashion or already picked AND make jams, fruit/herb spreads, wines, syrups, fruit leather, dried fruits and fruit-filled tarts and muffins.

Gourds and Pumpkins: Sell dried gourds to crafters AND gourd birdhouses, loofas, bowls, planters. Pumpkins are a fall favorite for everyone.

Popcorn: Sell as shelled popcorn AND sell on the cob as a novelty and in gift baskets.

Popcorn is easy to grow

Now let's move on to how and where you market your value-added products. As you read the following tips and suggestions, I want you to remember this…

No one has as much to gain or lose as you do from selling your products. This means no one should work harder at it than you do. – Darla Noble

Farmer's markets and craft shows/bazaars are usually the first place people go to sell their value-added products—and for good reason. Booth space is cheap, permits and food regulations are more relaxed because of the 'assumed risk' of buying from an open market and it's a fun and easy way to network with people who are doing what you are doing.

If yours isn't a product you can sell, per se, at a farmer's market (garden space, building rental, etc.) you can set up a booth with fliers, pictures and business cards to hand out.

Your next logical sales venue is online sales. There are a number of outlets for value-added agriculturists online. One of the best online stores available is Local Harvest (www.localharvest.org). Local Harvest makes it easy for sellers to set up their own virtual store and track sales for a minima commission. You don't have to have large amounts of inventory on hand and get world-wide exposure.

Etsy (www.etsy.com) is an option for those not selling food items. NOTE: Teas and spices can be sold. Again, you set up your own store, keep track of your own sales and Etsy takes only a small commission.

Amazon allows you to sell through their site (www.amazon.com) as an individual seller. They keep only $1 of every sale you make.

Social media sites (Facebook, Twitter, Google +, for instance) allow you to create a page for your value-added venture for free. You have the potential to reach literally millions of people this way for nothing more than two to three hours a week spent networking on these sites.

A free website. Webstarts.com allows you to build your own website using its free templates and backgrounds. Yes, you can upgrade which will require you to pay a fee, but even if you do decide to do that, the fees are minimal.

The internet is one of your most valuable sales/marketing tools. Don't ignore the tremendous power it has to help you reach the masses.

Aside from farmer's markets and the internet, you also need to make the following a part of your value-added program.
Business cards. Vista Print (www.vistaprint.com) offers them for very little cost.

Fliers. You need something to physically put in people's hands when they pass by your booth at a market or craft fair. You should also ask local/regional merchants to display your business card and flier on their bulletin boards, check-out counters, etc.

Mailing list. Create a mailing list by taking names/email addresses given by potential customers when they sign up to win free products from you. Send an email at least twice a month letting these people know what you have for sale, where you are selling and offer tips/suggestions and trivia about your products.

Sell wholesale. Selling wholesale means your markup will be less but you can usually sell in greater quantities when you sell wholesale so overall, it's a good thing. Selling wholesale also cuts your marketing costs (somewhat) and decreases the time you have invested in selling.

Join pertinent organizations. Being part of groups and organizations which focus on agriculture in general and your products/services specifically are an invaluable resource. Not only do they provide you with opportunities to learn how to enhance your value-added program, but they can be a great source of marketing and advertising. To find these organizations, search online for groups near you.

One excellent example of this is the *AgriMissouri* program under the MO Dept. of Agriculture. This program provides marketing tools and resources for Missouri farmers (small and large) as well as outlets for sales and an above-average amount of exposure. The fees are nominal and you get far more than you pay for. Missouri isn't the only state to offer such a program, so check with your state's dept. of agriculture to see what, if anything, they offer along these lines.

Don't just join in name only, though. Participate. Attend meetings. Work on a committee or two. Work for the good of the organization and not just your own interest. In doing so, you meet people, develop relationships, make a name for yourself and will be remembered by others when they are looking for products and services.

Spread the word. This isn't *Field of Dreams*. People are not going to buy just because you have something to buy. You have to let them know a) it's available b) they need it. Depending on what your product or service is, consider making contact with the following people and businesses in order to add value to what you grow/produce:

Professional offices such as banks, insurance companies, real estate offices and law offices are always in the market for unique client gifts, employee gifts and things to enhance their offices.

Florists, photographers and other wedding-oriented businesses are great to network with for a wedding venue.

Preschools, elementary schools, FFA, 4-H and scout troops will be delighted to hear about your farm tour experiences.

Your county extension office will be helpful in passing the word along about what you have to offer, so don't be afraid to let them know.

Local, regional and state tourism agencies will gladly make your brochures available to anyone who stops by looking for things to do.

Locally-owned gift shops, hospital gift shops and upscale boutiques will quite possibly be interested in carrying locally crafted items.

Seed-starting kits sold in gift stores add value to greenhouse owners and gardeners.

The mom and pop cafes are potential wholesale clients for your special sauce or relish, herbal breads or fresh fruit from your trees for their pies.

Health food stores are excited when they get the chance to offer home-grown specialty produce or eggs. They will also be glad to advertise the fact that you sell animals for meat (as long as they are raised naturally).

Churches and youth organizations are always looking for a place to have a hayride, have a party or learn something new.

Are you beginning to get excited? Are your creative juices flowing? I hope so, because value-added agriculture isn't just about making your hobby a profitable one. It's about having a great time doing it *and* promoting the farming industry; all of which are great things. After all—no farms…no food!

Chapter 5: Let's Talk Business

Now it's time for the technical stuff. I know, this isn't very fun or exciting, but it is important and essential. I'll try not to bore you by going into too much detail, but it would be irresponsible of me to fill your head with all sorts of ideas without telling you what you need to know to keep things legal and on the up and up.

NOTE: The information in this chapter is not meant to be taken as legal counsel. All information in this chapter is for informational purposes only. Each state, county and city's regulations differ in various ways. Therefore, each person should abide by the rules and regulations of the community in which they live regarding the production and selling of agricultural products.

Permits

Hobby farmers—even those with a value-added program—will generally not be required to have a business license for selling at a farmer's market or holiday craft markets. A business license will generally not be required when selling through sites such as Etsy or Local Harvest, either.

Depending on what you are selling, you may be required to have a business license in order to sell off the farm or to a business (retail or wholesale). Produce, livestock, livestock products and flowers would not fall under this 'blanket' because they are in their natural state. Once they have been processed, though, some places might not consider these things to be farm products any longer. You will need to find out what your state's laws are on this.

For those selling food products that have been cooked or processed in any way, most counties do not require these items to be prepared in a commercial kitchen which meets the standards of the local health department AS LONG AS you sell at a farmer's market. Buying and selling these items at a farmer's market is recognized as being a situation that carries what is called an 'assumable risk'. It is assumed that buyers know they are buying from local farmers/growers rather than from a commercial business. If you wish to sell to local businesses or out of a store, however, you will need to abide by all health department codes and requirements.

Cooking, processing and preparing food items to be sold online is a situation in which there is much room for interpretation of what is and isn't required or permitted. While the place you live may require these products to be made in a commercial kitchen to be sold in your area, the place you are selling to may not require this. It is something you need to put some serious thought and research into before beginning and make the choices that will be safe, healthy, efficient and effective for you.

If you wish to purchase items wholesale (tax free) for the purpose of producing and growing, you will need to obtain a tax-exempt status and identification number from your state's department of revenue. In turn, however, you will need to charge sales tax to your customers and pay that tax to the dept. of revenue.

Insurance and other legalities
If you are going to welcome visitors onto your property for parties, farm tours or other events or allow people onto your property for the purpose of growing their own garden or picking produce, you MUST have sufficient liability insurance coverage to protect yourself from being sued in case of an accident. You will also need to have signs posted stating you are not responsible for accidents as well as signs posted stating the rules of behavior.

Many agri-tourism (farm tour) venues require people to sign a waiver when visiting. This is usually reserved for school groups and such, but it is something to take into consideration. In place of a waiver, some farmers go over the rules with visitors and have them sign off on having heard them.

Liability protection is a must when conducting business of any kind on your premises

If your value-added program really takes off or is one in which there is a greater risk of liability, you might want to consider forming an LLC (Limited Liability Corporation). An LLC removes the possibility of someone suing you on a personal level. In other words, if a child falls and breaks their arm while at your farm and the parents decided to try to sue you, they could sue the business, but they couldn't touch your personal assets.

Marketing

We've already talked briefly about marketing and advertising, but let's take a closer look at how to make it optimally for you.
I cannot emphasize the importance of social media; Twitter and Facebook in particularly. Both are free and reach the masses.

You can set up a specific page for your business (Example: Generation 5 Farm Sheep, Greenhouse & Farm Tours) and invite your friends to 'like' it as well as hundreds of others who have pages that relate to yours in any way. Twitter limits the length of your message and will 'scold' you for posting a link to your page too many times in one setting, but inviting people to 'follow back' and 'like' your Facebook page is never a problem.

Even though we live in the world of technology don't ever underestimate the importance of print advertising. This is especially true when it comes to reaching out to schools and youth or civic organizations. They like something they can hold in their hand to see the 'who, what, where, when' and so forth. Business cards are also good reminders to people about who you are and what you have to offer.

Generally speaking, newspapers are not the best form of advertising these days. More and more people get their news online, on the radio and on television. So unless you are catering to an older population which still likes to read their news, spend your marketing money elsewhere.

Signage sells! A magnetic sign on your car door will generate lots of attention. I've seen more than a few people write down a phone number while sitting at a stoplight or take a picture of a sign on a car with their phone for future reference. T-shirts with your logo and contact info are also a great idea as are ink pens to hand out at farmer's markets and other events. These items aren't too terribly expensive, get results and are tax deductible.

Taxes

Speaking of taxes…

Value-added agriculture ventures are a form of farming. In other words, if what you are doing involves growing and producing food, livestock, or other ground crops, you are farming. And if you have income you need to file a Schedule F with your individual tax return.

The Schedule F is not difficult to do as long as you have documentation for the expenses you claim and income you report. Keeping track of both income and expenses is not only for tax purposes, though, but also as an indicator as to how well you are doing.

Most people aren't sure of what they can and cannot deduct on their taxes. The instruction booklet (print or online) gives a thorough list, but here are just a few you some people overlook that you need to be sure to take advantage of:

*Mileage to and from delivering goods and selling at markets, etc.

*A percentage of your internet fees if you do any business online

*Water and utility charges in the barn or other outbuildings on your farm used for agricultural purposes

*Advertising and marketing expenses

*Meals eaten at events you attend for agricultural purposes (1/2 the cost of the meal is the allowable expense)

*A percentage of your cell phone or home phone fees if used to conduct business

*Feed and supplies for livestock

*Seed, fertilizer and tools used for food or crop production

*Insurance premiums for liability protection

*Dues paid to organizations for membership

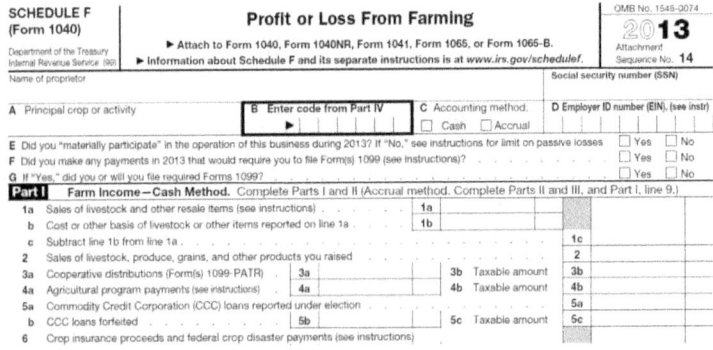

IRS Schedule F

Financial records

How much money you spend and how much you take in; that's the basis for your financial record keeping. It really is that simple. I advise you to break it down a bit more, but essentially all you really need to know is how much you are spending on your venture vs. how much money it is bringing in.

Keep track of all expenses—from the biggest to the tiniest seed, new baby chick or even a bottle of bug spray for the plants in your greenhouse. They all matter and they all affect the bottom line.

Keeping track of income is equally important; not only the amount of income, but where it derived from. Knowing where it comes from isn't so much for tax purposes as it is for good business purposes. Knowing where you make the most sales tells you where to concentrate your efforts.

EXCEL spread sheets are easy and versatile for any type of project and are available to any Microsoft user. If you aren't into computers, though, there is nothing wrong with a ledger you actually write in…with a pen…and your hand.

The method you use for keeping track of your financial income and expenses isn't nearly as important as the fact that you just *do* it. Otherwise you will not know if you are making a profit or not.

Chapter 6: Helpful Resources

The following is a list of websites that provide information, resources and outlets for a successful value-added agricultural venture you can be proud of.

www.localharvest.org
http://sdvalueadded.coop/
http://fyi.uwex.edu/aic/2012/02/10/usda-value-added-producer-grants-turning-great-ideas-into-sustainable-business/
http://ag.arizona.edu/arec/va/valaddopp&const.html
http://agrimissouri.com/
http://www.harrisseeds.com/
https://www.pinterest.com/
www.tendingmygarden.com
www.seedman.com
http://www.northamericanfarmer.com/
http://www.ansc.purdue.edu/SH/Resources/Publications/Extension%20Articles.htm
https://ag.purdue.edu/ansc/Pages/default.aspx

Conclusion

"I can take a joke as well as the next person, but I've always taken offense to the one about the farmer who, when asked what he's going to do with the money he inherited, says he's going to farm 'til it's all gone. Take it from this 4th generation farm gal; I'm not in it to lose money. And you don't have to be either!"
–Darla Noble

Farming-whether it's a hobby, supplemental income, or your sole livelihood--can (and should) be profitable. Value-added agriculture is one way to make that happen. I hope what you have read has served to enlighten, inform and encourage you to become an even greater part of the agricultural community—the people that literally keep us alive.

Author Bio

Darla Noble is a native of mid-Missouri where she lives with her husband of thirty-three years, John. Darla's love of writing began in the fourth grade; after meeting up and coming children's author, Judy Blume,
who, by the way, autographed Darla's copy of "Are you there, God...it's me, Margaret".

Darla's love for writing and family makes her work sought after in the Christian market, parenting and family resources and ghostwriting for educators and inspirational speakers.

Health Learning Series

- THE MAGIC OF GOOSEBERRIES FOR HEALTH AND BEAUTY
- THE MAGIC OF YOGURT FOR COOKING AND BEAUTY
- THE MAGIC OF LEMONS — USING LEMONS FOR HEALTH AND BEAUTY
- THE MAGIC OF CHILLIES FOR COOKING AND HEALING
- THE MAGIC OF ONIONS — ONIONS IN CUISINE TO CURE AND TO HEAL
- THE MAGIC OF RADISHES TO CURE AND TO HEAL
- THE MAGIC OF CARROTS TO CURE AND TO HEAL
- THE HEALTH BENEFITS OF OREGANO FOR COOKING AND HEALTH
- The Magic of MARIGOLDS — Marigolds for Health And Beauty
- THE HEALTH BENEFITS OF CINNAMON
- THE MAGIC OF COCONUTS FOR COOKING & HEALTH
- THE MAGIC OF CLOVES FOR HEALING AND COOKING
- THE MAGIC OF ASAFETIDA FOR COOKING AND HEALING
- THE MAGIC OF NEEM — MARGOSA TO HEAL
- THE MAGIC OF SALT TO HEAL AND FOR BEAUTY
- THE HEALTH BENEFITS OF POMEGRANATES FOR HEALTH AND BEAUTY
- THE MAGIC OF DRY FRUIT AND SPICES — REMEDIES AND RECIPES
- THE HEALTH BENEFITS OF TURMERIC CURCUMIN FOR COOKING AND HEALTH
- THE MAGIC OF ALOE VERA
- THE MAGIC OF VEGETABLES — ANCIENT HEALING REMEDIES AND TIPS
- THE HEALTH BENEFITS OF ROSEMARY FOR COOKING AND HEALTH
- THE MAGIC OF PEPPER & PEPPERCORNS FOR COOKING & HEALING
- THE MAGIC OF MILK, BUTTER AND CHEESE FOR COOKING & HEALING
- THE MAGIC OF CARDAMOMS FOR COOKING AND HEALTH
- THE HEALTH BENEFITS OF BLACK CUMIN FOR COOKING AND HEALTH
- THE MAGIC OF BASIL-TULSI TO HEAL NATURALLY
- THE MAGIC OF SPICES FOR HEALTH AND CUISINE
- THE MAGIC OF ROSES FOR COOKING AND BEAUTY
- The Miraculous Healing Powers of GINGER
- The Miracle of HONEY

Amazing Animal Book Series

Learn To Draw Series

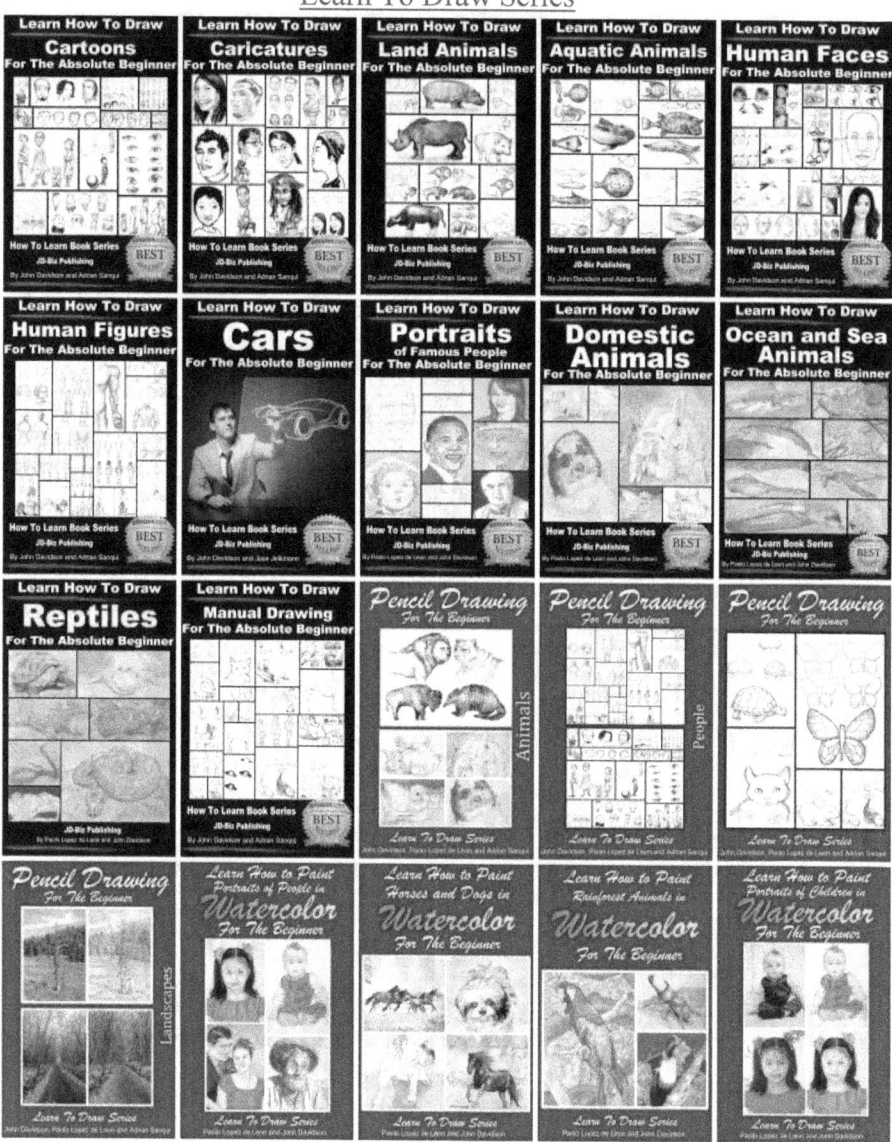

How to Build and Plan Books

Entrepreneur Book Series

This book is published by

JD-Biz Corp

P O Box 374

Mendon, Utah 84325

http://www.jd-biz.com/

Read more books from

www.ingramcontent.com/pod-product-compliance
Lightning Source LLC
Chambersburg PA
CBHW070716180526
45167CB00004B/1491